The Story of Our Numbers

The History of Arabic Numerals

Zelda King

PowerMath™

The Rosen Publishing Group's
PowerKids Press™
New York

Published in 2004 by The Rosen Publishing Group, Inc.
29 East 21st Street, New York, NY 10010

Book Design: Haley Wilson

Photo Credits: p. 4 © Mimmo Jodice/Corbis; p. 6 © Richard T. Nowitz/Corbis; p. 6 (Colosseum) ©
ML Sinibald/Corbis; p. 10 © Arne Hodalic/Corbis; p. 14 © Carmen Redondo/Corbis; p. 16 ©
Adam Woolfitt/Corbis; pp. 18, 20 © Bettmann/Corbis; p. 21 © Todd Gipstein/Corbis.

Library of Congress Cataloging-in-Publication Data

King, Zelda.
  The story of our numbers : the history of Arabic numerals / Zelda
King.
      v. cm. — (PowerMath)
Contents: The numerals of ancient Rome — The birth of our numerals —
Hindu numerals spread to Arab lands — Hindu-Arabic numerals reach
Europe — The death of Roman numerals.
  ISBN 0-8239-8965-8 (lib. bdg.)
  ISBN 0-8239-8870-8 (pbk.)
  6-pack ISBN: 0-8239-7378-6
  1. Numeration, Arabic—History—Juvenile literature. 2. Roman
numerals—Juvenile literature. [1. Numerals. 2. Roman numerals. 3.
Number systems.] I. Title. II. Series.
  QA141.3.K56 2004
  513.5—dc21
                                    2002156664

Manufactured in the United States of America

Contents

Rome ● ITALY

Mediterranean Sea

The Numerals of Ancient Rome

Imagine that you lived in ancient Rome 2,000 years ago. You would have studied math just like students do today. However, you would have used Roman **numerals** rather than the numerals we use today.

Romans formed all their numbers from only 7 numerals. For numbers consisting of more than 1 numeral, the usual rule was that you added the numerals together to get the value of the number. For example,

Roman Numeral	Our Number
I	1
V	5
X	10
L	50
C	100
D	500
M	1,000

III = 1 + 1 + 1 = 3, **XV = 10 + 5 = 15**, **CC = 100 + 100 = 200**, and so on. However, in a number such as **IV**, where the numeral on the left has a lower value than the numeral on the right, you subtracted the left numeral from the right one. So **IV = 5 − 1 = 4**. In the same way, **IX = 10 − 1 = 9**, **XC = 100 − 10 = 90**, and so on.

Roman children were taught at home until they were 12 years old. Then boys went to school for further study. Girls stayed at home where their mothers taught them the skills they needed to run a home well.

Colosseum

To see what it was like to do math with Roman numerals, let's try a few problems.

V + V = X   XV + IV = XIX
5 + 5 = 10   15 + 4 = 19
XXX − VII = XXIII   XV − VII = VIII
30 − 7 = 23   15 − 7 = 8

Even simple addition and subtraction problems like these can be confusing with Roman numerals. Imagine trying to multiply with Roman numerals!

III x V = XV   V x V = XXV   X x IV = XL
3 x 5 = 15   5 x 5 = 25   10 x 4 = 40

After this look at ancient Roman math, you're probably very glad that we don't use Roman numerals today! However, you may also be wondering where the numerals we do use came from. This book will tell the story of our numbers.

Ancient Romans were famous for their buildings, like the Roman Colosseum, and their roads. Planning and making these buildings and roads required great skills in math.

1 2 3 4 5 6 7 8 9 0

Ancient Nagari numerals

Nagari numerals around 11th century A.D.

8

The Birth of Our Numerals

You may have heard people call the numerals we use "Arabic" numerals. This is because the people of Europe learned them from the Arabs. However, **Muslim** Arab **mathematicians** had learned the numerals from **Hindu** mathematicians in India. A better name for our numbers is "Hindu-Arabic" numerals.

The story of our numerals began in India around 250 B.C., with the invention of numbers called the Brahmi (BRAH-mee) numerals. These Brahmi numerals were the earliest form of the numerals we use today, but they did not look much like our numerals.

Brahmi numerals around 1st century A.D.

1	2	3	4	5	6	7	8	9
−	=	≡	+	ɦ	ϛ	ʔ	Ϥ	ʔ

Brahmi numerals changed slowly over time. By around 650 A.D., the numbers had taken new forms and were known as the Nagari (NAH-guh-ree) numerals. Nagari numerals included a zero, which the Brahmi numerals did not have.

CHINA

Great Himalaya Range

New
Delhi★

Ganges

I N D I A

Arabian Sea

Calcutta

Bay
of
Bengal

Bombay

Madras

SRI
LANKA

10

Aryabhata probably lived
and worked in a city near
the Ganges River in
northern India.

In addition to the actual forms of the numerals, our number system has another very important feature. It is a place-value number system. Neither the Brahmi numerals nor the Nagari numerals used a place-value number system. However, just like the shapes of our numerals, the place-value system was invented in India.

A famous Hindu mathematician named Aryabhata (ar-yuhb-HUT-uh) invented a place-value system when he was only 23 years old. Around 500 A.D., Aryabhata wrote a book on **astronomy**. The book was written in verse, like a long poem. It included a history of Hindu math up to that time and explains Aryabhata's place-value system. The book, however, did not use the Brahmi numerals. Aryabhata still followed a very ancient system that used letters from the Hindu alphabet to stand for numbers.

No one knows exactly when the Hindu place-value system and the Brahmi or Nagari numerals were first used together. From written records, we do know that the place-value system and Nagari numerals were being used together by around 875 A.D.

Arab empire around 900 A.D.

Spain

Black Sea

Caspian Sea

Mediterranean Sea

North Africa

Middle East

Asia

Red Sea

Arabian Sea

Al-Khwarizmi's book helped to spread knowledge of the Hindu system throughout the Arab world. At the time he wrote his book, Arabs ruled the Middle East, North Africa, most of Spain, and part of western Asia.

Hindu Numerals Spread to Arab Lands

In 776 A.D., an Indian book on astronomy was brought to an Arab ruler in what is now the country of Iraq. The ruler had the book **translated** into Arabic so that Arab mathematicians could learn the Hindu number system. Then, around 825 A.D., a famous Arab mathematician named al-Khwarizmi (al–hwar-REEZ-me) wrote a book on the Hindu number system.

Arab mathematicians used the name "*ghubar*" (ROO-bar), or "dust," for the Hindu numerals. This was because they worked out math problems by writing in dust spread on a wooden board. They could then erase and make changes as they solved the problems, much like we erase and make changes on a chalkboard today.

Some people did not like to use dust boards because they were messy. Around 950 A.D., an Arab mathematician named al-Uqlidisi (al–oo-KLEE-dee-see) invented a way to write out the steps in solving a math problem using pen and paper. He included his new method in a book he wrote about Hindu numerals and the place-value system.

The university where al-Banna taught still exists today in Fez, Morocco. It is one of the oldest universities in the world!

By around 1300 A.D., Hindu-Arabic numerals had begun to look much more like the numerals we use today. These numerals appeared in a book written by an Arab mathematician named al-Banna (al–BAN-nah).

al-Banna's numerals

1	2	3	4	5	6	7	8	9
٢	٢	٤	٢	٢	٢	٢	٨	٢

Al-Banna was a famous math teacher at the university in Fez, in what is now the country of Morocco. Fez was a wealthy city with a royal palace as well as an important learning center. Students came to Fez from distant places in order to study with al-Banna. Books written by al-Banna helped to spread his ideas as well as the ideas he had learned from other mathematicians.

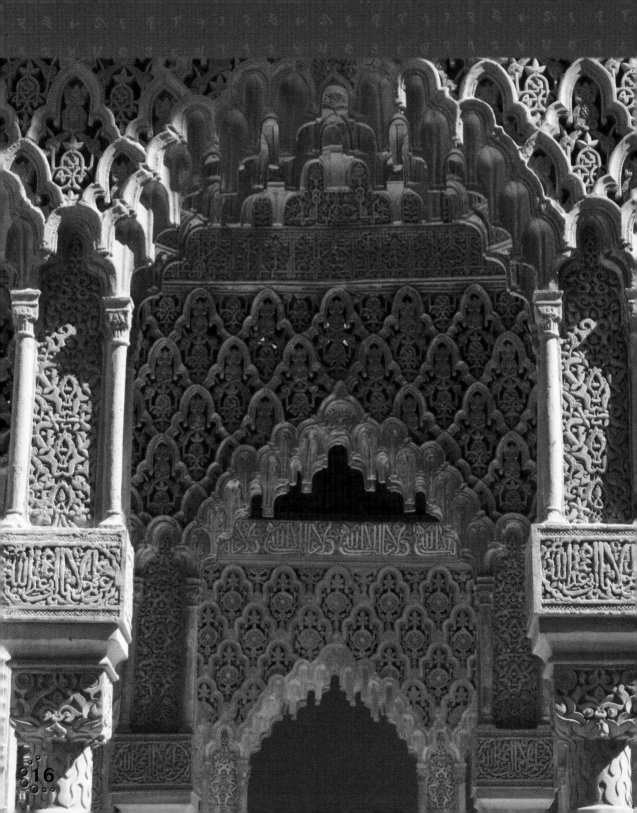

16

Hindu-Arabic Numerals Reach Europe

In 711 A.D., Muslim Arabs gained control of most of Spain, where they established great centers of power and learning. Christian Europeans kept control of a small area in northern Spain. However, they recognized the Arabs' great accomplishments in art and knowledge, and learned many things from them. A **monk** living in northern Spain included Hindu-Arabic numerals in a book he wrote in 976 A.D.

A few other educated Europeans also saw the advantages of the Hindu-Arabic number system. A French monk named Gerbert wrote a book that explained the system around 980 A.D. Gerbert had learned about the Hindu-Arabic number system while he was studying in Spain between 967 A.D. and 969 A.D.

Most Europeans were slow to realize the advantages of Hindu-Arabic numerals. Merchants continued to use the Roman numerals they had used for centuries. Gradually, European mathematicians became interested in Hindu-Arabic numerals and began to study works written by men like al-Khwarizmi.

This picture shows part of the beautiful Muslim palace called the Alhambra in Spain.

"Fibonacci" was actually a nickname that meant "son of the Bonacci family." His real name was Leonardo Pisano. Sometimes Fibonacci also used the name "Bigollo," an Italian word that meant "traveler."

An Italian merchant's son named Leonardo Fibonacci (fih-buh-NAH-chee) wrote a book on the Hindu-Arabic number system in 1202 A.D. Fibonacci's father was in charge of an Italian trading colony in the Arab area of North Africa. To train Fibonacci in the skills he would need to be a merchant, his father arranged for him to learn the Hindu-Arabic number system from Arab merchants.

Fibonacci quickly recognized the benefits of the Hindu-Arabic system. After he returned to Italy around 1200 A.D., he began work on a book that would explain the benefits to other Europeans. Fibonacci had a special interest in the ways the Hindu-Arabic system could help people solve real-life math problems. His book presented the kinds of math problems that merchants had to deal with every day.

Fibonacci's book became so popular that it was copied many times by other European mathematicians. However, most Europeans continued to use Roman numerals. In fact, some places in Italy made it illegal to use anything but Roman numerals! Government officials feared that some people might try to cheat others by changing the forms of Hindu-Arabic numerals.

Printers arranged type in the tray, then spread ink over it. The tray and a sheet of paper were put in the printing press together, and the page was printed.

The Death of Roman Numerals

What finally caused Europeans to accept the Hindu-Arabic numerals? It was the invention of the **printing press**. For centuries, all books in Europe had been written by hand. Then around 1450 A.D., Europeans learned how to print books. This allowed them to make hundreds of copies of a book much more quickly than copies could be made by hand.

Printers preferred Hindu-Arabic numerals because they made the printing process faster and easier. We can see why this was so if we look at the printing method used. For each page, the printer arranged metal **type** in a tray. Each letter and number was on a separate block, and the letter or number was raised, like the picture on a rubber stamp today.

Hindu-Arabic numerals made the printing process faster because the numbers usually required fewer digits than numbers written with Roman numerals. For example, it takes 2 digits to write the number of this page using Hindu-Arabic numerals: 22. Using Roman numerals, it would take 4 digits: XXII. By using fewer blocks of metal type, printers saved a little time on each page and a lot of time on a whole book. By around 1550 A.D., Hindu-Arabic numerals had taken the place of Roman numerals throughout Europe.

Today you sometimes see Roman numerals on a clock or in a book, but we use Hindu-Arabic numerals for almost everything. Next time you work on a math problem, think about how much harder it would be if you had to use Roman numerals. You'll be glad we have Hindu-Arabic numerals!

Glossary

astronomy (uh-STRAH-nuh-mee) The study of the Sun, Moon, stars, and other space objects.

Hindu (HIN-doo) Someone who practices the faith of Hinduism.

mathematician (math-muh-TIH-shun) Someone who is an authority on math.

monk (MUHNK) A man who has promised to live according to rules made by the church and who lives in a community with other men who have made the same promise.

Muslim (MUHZ-luhm) Someone who practices the faith of Islam.

numeral (NOO-muh-ruhl) A sign that stands for a number.

printing press (PRIN-ting PRESS) A machine used to print things like books and posters.

translate (TRANS-layt) To take something written in one language and write it in another language.

type (TYPE) Metal blocks with raised letters or numbers that can be put together and inked to print something on paper.

Index